9.95

Deserts

by Anna O'Mara

Bridgestone Books

an Imprint of Capstone Press

Facts about Deserts

- Deserts are found on all the continents except Europe.
- The hottest temperature on earth was recorded in the Sahara Desert in Africa.
- Ten inches (25 centimeters) of rain or less fall each year in most deserts.
- The hottest place in North America is Death Valley in the Mojave Desert.

Bridgestone Books are published by Capstone Press • 818 North Willow Street, Mankato, Minnesota 56001
Copyright © 1996 by Capstone Press • All rights reserved • Printed in the United States of America

Library of Congress Cataloging-in-Publication Data
O'Mara, Anna.
    Deserts/Anna O'Mara.
    p. cm.
    Includes bibliographical references and index.
    Summary: Provides basic information about the conditions, terrain, plants, and animals found in the
    major deserts of the world.
    ISBN 1-56065-338-8
    1. Desert ecology--Juvenile literature. 2. Deserts--Juvenile literature. [1. Deserts.] I. Title.
QH541.5.D40535 1996
574.5'2652--dc20

                                        95-43240
                                        CIP
                                        AC

Photo credits
J.P. Rowan: cover, 4, 10, 16, 20. Jean Buldain: 8. Scenics, ETC.: 12, 18
Sage Productions/Sharon K. Snedden: 14.

Table of Contents

Words in **boldface** type in the text are defined in the Words to Know section in the back of this book.

Dry Deserts

Deserts are dry land areas. Most people think that deserts are full of sand. But only about one desert in five is sandy. Small stones cover most deserts.

A desert does not get much rain. Ten inches (25 centimeters) or less of rain fall each year in most deserts. Some deserts have no rain at all for several years. This makes deserts very dry.

Hills and mountains surround deserts. The mountain valleys have dry streams. People in North America use a Spanish word for these dry streams. They call them **arroyos**. People in Africa and the Middle East call them **wadis**.

The arroyos and wadis are dry most of the year. But water sometimes rushes down them. The desert quickly swallows up the water from these flash floods.

The organ-pipe cactus grows in the Sonoran Desert in Arizona.

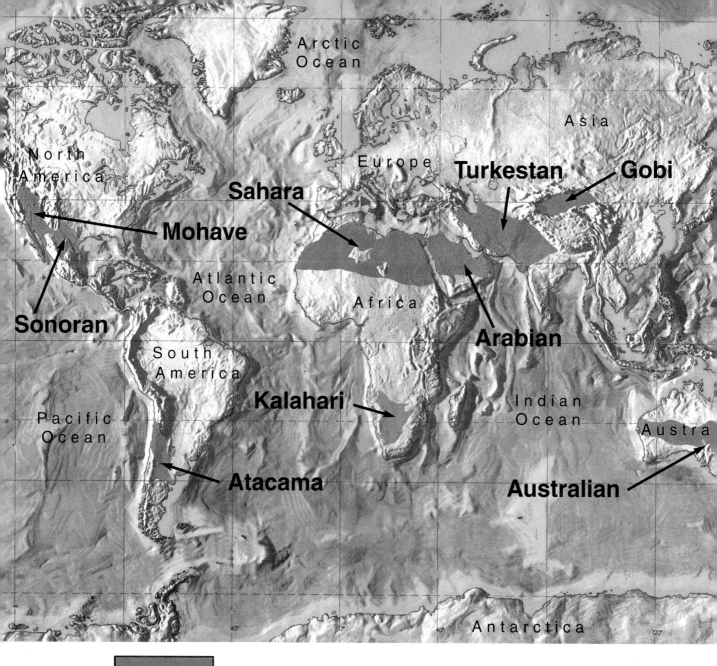

Deserts of the World

Deserts of the World

The Sahara Desert is the biggest desert in the world. It reaches 3,200 miles (5,120 kilometers) across the top of North Africa. It covers nearly one-third of Africa. The Sahara Desert is the size of the entire United States.

The next biggest desert is the Australian Desert. It covers nearly half of the continent of Australia.

The Arabian Desert is a large desert in Saudi Arabia.

Other large deserts are the Gobi Desert in Mongolia and China and the Kalahari Desert in southern Africa.

In North America, deserts are found between the Rocky Mountains and the Sierra Nevada mountains.

High Temperatures

Deserts are dry and hot. The hottest air temperature on earth was recorded in the Sahara Desert. It was 136 degrees Fahrenheit (58 degrees Celsius).

The hottest place in North America is Death Valley. It is in the Mojave Desert of California. The summer temperatures there are sometimes above 120 degrees Fahrenheit (49 degrees Celsius).

During the day, no clouds guard the desert from the rays of the sun. At night, the heat of the day escapes. The temperature can drop to 45 degrees Fahrenheit (7 degrees Celsius) or lower.

In some deserts, like the Gobi Desert in Mongolia and China, winters are long and cold. At night the temperature falls as low as 32 degrees Fahrenheit (zero degrees Celsius). During the winter, snow falls in the Gobi Desert.

Death Valley is the hottest place in North America.

Rain, Floods, and Flowers

Many deserts get less than one inch (2.5 centimeters) of rain each year. Sometimes, after no moisture for many years, rain will fall.

Heavy rains can cause flooding in the desert. The rain runs off the hills and mountains and into the dry beds of the streams. The rain water quickly becomes a flood. Water roars down into the valleys.

The water floods out of the valleys and across the desert. The dry rocky ground drinks some of the water. The rest of the water **evaporates.**

The rain and floods break open seeds in the ground. These seeds quickly sprout, grow, and bloom. A carpet of colorful flowers covers the desert for a few days.

Before the flowers die, they produce seeds. The seeds fall into the soil and wait for the next rain. Then they, too, will sprout, grow, and bloom.

The claret-cup cactus grows in the desert regions of Texas.

Desert Plants

Many plants grow in the desert. They can live without much water.

Plants like the **saguaro** cactus store water in their stems and branches. The saguaro grows in the Sonoran Desert of the United States and Mexico.

When rain falls, the saguaro's roots drink as much water as possible. The stems and branches fill up with water. The cactus lives on this water during dry weather. Its sides contract slowly as it drinks this water. A saguaro cactus can live to be 200 years old.

Many cactus plants have waxy skins. This keeps water inside from evaporating. The plants' sharp needles keep animals from eating them or taking their water.

Joshua trees are yucca plants that grow in North American deserts and semidesert areas. They can grow as tall as 40 feet (12 meters). Joshua trees have white flowers that smell sweet at night.

The saguaro cactus grows in the Sonoran Desert.

People of the Desert

Not many people live in the desert. Those who do live there are often nomads. They are people who move from place to place. They search for food and water.

The nomads look for an oasis. It is found where water comes up through the ground. An oasis has trees and plants growing around a pool.

Nomads herd sheep, goats, cattle, and camels. Nomads and their animals travel many miles to drink at an oasis.

In Africa and Asia, nomads live in tents. They dress in white robes that cover their bodies. The robes guard against the sun and the sand.

Some desert people have farms. They raise their livestock in one place. In North America, many Native Americans live in the desert. They often have **adobe** houses. The thick adobe walls keep the houses cool during the day.

Nomads who live in the desert get water from an oasis.

Bugs and Reptiles

Many bugs and reptiles live in the desert. The bugs have long legs. These keep the bugs' bodies above the hot desert floor.

Beetles have a thick outer covering. It keeps water inside their bodies. Tarantulas with poisonous teeth catch the beetles. Scorpions with poisonous stingers also attack them.

The beetles dig into the ground to escape the bugs and the heat. The desert surface is hot. A few inches beneath the surface, however, the sand is much cooler. There the beetles are safe.

The Gila monster is a poisonous lizard. This reptile lives in the deserts of North America. Its scales keep it from drying out in the desert heat.

The rattlesnake, another reptile, also has scaly skin. The scales keep the snake from losing water.

The wolf spider lives in the deserts of Arizona.

Desert Animals

Desert animals are suited to their home. The kangaroo rat, the jack rabbit, the coyote, and the camel are suited in different ways.

The kangaroo rat lives in deserts in North America. It lives in cool holes under the desert floor. At night it is not so hot in the desert. The kangaroo rat comes out to eat.

Some animals, like the jack rabbit, do not live beneath the desert floor. They lie in the shade of desert plants. The jack rabbit stays cool. Its body heat escapes through its large ears.

An Arabian camel can go for 17 days without water. It walks to an oasis. There it drinks as much as 25 gallons (95 liters) of water.

The camel can go without food for many days, too. The camel stores fat in the hump on its back. When the camel cannot find food to eat, it lives on the fat in this hump.

Coyotes live throughout most of North America, including the deserts of the southwest United States.

Spread of Deserts

There are more deserts on earth now than in the past. Deserts are growing. New deserts are forming. This happens when animals eat too much grass near deserts. The grass never grows back.

Wind and rain wear away the soil. The wind blows the soil away. Then a desert forms. The change of fruitful land into a desert is called **desertification**.

Mining and tree-cutting also cause desertification. It is a big problem.

People must take care of the land. Then the deserts will stop spreading.

The Gypsum Sand Dunes are in New Mexico.

Hands On: Make a Piece of Desert Art

Native Americans who live in the desert make sand paintings. The Navajo make them to help cure illnesses and to bring rain.

You can make your own sand painting.

You will need

paper, a flat piece of sandpaper, a pencil, spray varnish, colored sand (you can buy it at a hobby store or color your own using spices like paprika, cinnamon, turmeric, or dry mustard)

How to do it

1. Sketch a design on paper. Then draw the design with a pencil on sandpaper.
2. Use your fingers, or a spoon, or a small funnel to sprinkle the sand along the lines of your design.
3. Use spray varnish to save the drawing.

Words to Know

adobe—a brick of sun-dried earth and straw

arroyo—a dry or flowing stream in a North American desert

desertification—the change of fruitful land into a desert

evaporate—to disappear into a vapor

saguaro (sah-WAR-oh)—a kind of cactus

wadi—a dry or flowing stream in a desert in Africa or Asia

Read More

Bender, Lionel. *The Story of the Earth: Desert.* New York: Franklin Watts, 1989.

Stephen, Richard. *Deserts.* Mahwah, N.J.: Troll Associates, 1990.

Taylor, Barbara. *Desert Life.* New York: Dorling Kindersley, 1992.

Wingfield, John C. *Deserts of America.* Milwaukee, Wis.: Raintree Publishers, 1989.

Useful Addresses

Anza-Borrego Desert State Park
Box 299
Borrego Springs, CA 92004

Chihuahuan Desert Research Institute and Visitor Center
Box 1334
Alpine, TX 79831

Arizona Sonora Desert Museum
2021 N. Kinney Road
Tucson, AZ 85743

Desert Botanical Garden
1201 N. Galvin Parkway
Papago Park
Phoenix, AZ 85008

Index